Ranthambhore

— 10 DAYS —
IN THE TIGER FORTRESS

Ranthambhore
— 10 DAYS —
IN THE TIGER FORTRESS

Valmik Thapar

OXFORD
UNIVERSITY PRESS

OXFORD
UNIVERSITY PRESS

YMCA Library Building, Jai Singh Road, New Delhi 110 001

Oxford University Press is a department of the University of Oxford.
It furthers the University's objective of excellence in research, scholarship,
and education by publishing worldwide in

Oxford   New York
Auckland   Cape Town   Dar es Salaam   Hong Kong   Karachi
Kuala Lumpur   Madrid   Melbourne   Mexico City   Nairobi
New Delhi   Shanghai   Taipei   Toronto

With offices in
Argentina   Austria   Brazil   Chile   Czech Republic   France   Greece
Guatemala   Hungary   Italy   Japan   Poland   Portugal   Singapore
South Korea   Switzerland   Thailand   Turkey   Ukraine   Vietnam

Oxford is a registered trade mark of Oxford University Press
in the UK and in certain other countries.

Published in India by Oxford University Press, New Delhi

ISBN-13: 978-0-19-569946-3
ISBN-10: 0-19-569946-7

Typeset in Bodoni Bk BT 11/16
Designed by Ka Designs www.kadesigns.co.in
Printed in India by Thomson Press
Published by Oxford University Press
YMCA Library Building, Jai Singh Road, New Delhi 110 001

This book is for my son Hamir and my nephew Jaisal,
both of whom first experienced Ranthambhore's magical setting when
they were babes in arms. May they never forget this.

in acknowledgement ...

I believed that I would never do another book on tigers. But the magic of Ranthambhore enveloped and inspired me in March–April of 2008, and I was left with no other choice ... so came this book. For me it is a very special endeavour, for it was also the first time I watched my son use a camera and engage with the tigers and wildlife of Ranthambhore. What we saw was spectacular and took me down memory lane to those early days of 1975. Except for the end papers, the pictures are as they were taken on each of the days I was there.

All my pictures have been taken with the superb Lumex camera FZ50 and FZ7. So have Sanjna's. This camera, with its Leica lens, has the capacity to zoom from 28mm to 500mm. I also occasionally used the Sony Zeiss combination—the DSCH9—which was also a great pleasure. These digital cameras are wonderful and have opened a whole new world for capturing those very special wildlife moments.

I must thank my jeep passengers who at different times were on this journey with us—Amita for her excellent pictures, Jaisal and Anjali for some of theirs, and Raghu and Joanna for strengthening the visuals with their professional cameras. My wife Sanjna's pictures are, of course, always a delight.

Last but not least Hamir, who was not only fantastic in the jeep but whose pictures were a great first-time effort with a camera.

I must also thank R.S. Shekhavat, Deputy Director of Ranthambhore Tiger Reserve, for his very special assistance; Daulat Singh, Range Forest Officer, for his pictures of us in a jeep following two tigers; Shyamji for being just an excellent driver; and all my old friends in Ranthambhore in the forest staff and outside for keeping the magic alive.

I wish that those who dictate policy governing the natural world had a chance to witness wild tigers like I have had. That's when you understand what a tiger needs. On 1 November 2007, in a presentation I made to the Prime Minister and the National Board of Wildlife, this is the single fact I reiterated again and again: till you understand what a tiger needs you cannot save it. Sadly the government seldom understands the tiger's needs. Bureaucrats advise Chief Ministers and Prime Ministers without understanding the problem. Committees get constituted and task forces are created with people on board who have little knowledge of wild tigers. Luckily the shock of Sariska prevented this from happening in Ranthambhore. In fact the Chief Minister of

Rajasthan created a genuine group of experts to advise her. She listened and acted on their advice. Increasing protection in Ranthambhore combined with a hand-picked team of forest officers has done the trick. That's what we need all across India—strengthen and upgrade protection, and carefully choose the team that manages tiger reserves. The welfare of people has to be separated from the welfare of the tiger; they don't mix. It should also be mandatory for senior politicians who govern tiger affairs to spend a week in a year with them. Then we will see signs of success.

I must also specially thank my old tiger guru Fateh Singh Rathore, the original creator of all this magic. I thank Oxford University Press for agreeing to publish this book at lightning speed. The design team in Ka Designs lead by Anisha Heble and assisted by Radhika Chitalia and Mamta Somaiya have done a fantastic job. I would also like to thank Koustubh Sharma for all his help in organizing the pictures on the computer. I would like the readers to remember that this book is a completely authentic account of ten days spent with Ranthambhore's tigers.

Valmik Thapar

1 May 2008

a magical setting

For thirty-three years, since I first went to Ranthambhore, the setting of this unique wilderness with its massive fort, rolling hills, and lakes has mesmerized me. Its rich forest is dotted with the crumbling ruins of an ancient past. The fort sits like a citadel looking out at the splendid forest with its spectacular trees. Its history goes back a thousand years, and the great Mughal emperor Akbar waged war for its control. This amazing connection between its past and wild present captivated and enveloped me. I spent the best years of my life living in Jogi Mahal at the edge of the lake with the massive fortress of Ranthambhore looming behind. From the balcony of Jogi Mahal I have witnessed some fantastic encounters as tigers raced to the edges of the lake and even inside it in their efforts to catch deer. I have seen tigers fighting off the crocodiles which infest the waters of this lake. Enveloping Jogi Mahal in its folds is one of the largest and most magnificent banyan trees I have ever seen. It was under its branches that I spent endless days and nights in my quest to discover the secret life of tigers.

The Ranthambhore tigers are rewriting natural history amidst the memories of man. The palaces and ruins have been completely taken over by tigers. It is in this remarkable backdrop that our story is set.

introduction

This is my story and it happened at a moment of great turbulence for the tigers of India. In early 2008 the final estimation of tigers left in India done by the federal government came to the shocking figure of 1400. In a way all my fears at the turn of the century had been justified. In fact this is roughly the figure that I had predicted many years ago and which few had bothered to believe. Inept and inefficient governments had allowed the tiger to slide towards oblivion. I remember stating in my book *The Last Tiger* (2006) that only a miracle could save the tiger in India. In March 2008 I believed the tiger was steadily walking down the road to near extinction. All our governments—central and state—had failed to deal with the crisis. Just simple strategies of protecting the tiger's home or flushing out poachers from their jungle hideouts had been completely ignored. Thousands of tigers had died to service the needs of the Chinese for skin and bone for traditional medicine. Indian poachers had had a field day. And the political will of this nation had failed to keep wild tigers safe.

I had tried everything in my armoury to save this animal. It had been thirty-three years of effort but this March of 2008 must have been my gloomiest moment. I was enveloped in feelings of not just my own failure but this country's complete inability to govern and protect the natural world. Why was it so difficult for the government to act? Why did they not care? The solutions were simple, basic commonsensical measures. Were we now a country without the sense to save a unique wilderness with

its diverse species? Would global warming and climate change devastate this region? Would everything vanish under the economic onslaught of this strange century? While all these thoughts chased each other anarchically in my mind, I found that my son's ten-day Easter vacation was about to start. My wife was vanishing for a week for a theatre meet in interior Karnataka, and like a flash I jumped on a window of opportunity, deciding to vanish into my beautiful Ranthambhore with my son. Nearly 6, I felt he was now at an age to engage. Armed with his first camera ever, we caught an afternoon train to Ranthambhore. What follows is our story over nine days in Ranthambhore as it happened. On the last two days my wife joined us. I said in *The Last Tiger* that I would not do another book unless there was some kind of miracle in the life of the tiger. I believe now that we were witness to the magic of tigers in Ranthambhore between 22 and 31 March 2008. That is why I have put pen and picture to paper again. That is why I believe once again that the tiger can be saved through basic common sense and simple interventions. Ranthambhore proves it. This turned out to be one of my finest trips in thirty-three years, and it left me stunned and overwhelmed. While the tigers of Ranthambhore gave of their best, it was also a superb engagement for Hamir with the forest. And therefore this is a personal story of this remarkable journey with Ranthambhore's tigers over ten days. All this in a moment of severe gloom and desperation surrounding the tiger's future.

We arrived in Ranthambhore late one evening and as the next day was Holi we decided to stay home. You can be drenched with colour on your way in and out of the Park. Our home exists on about 10 acres of land that I bought twenty years ago when it was totally arid and had not a blade of grass on it. I planted with my own hands 4000 trees of sixty diverse species, many of them from seeds found in the Park. In this desert-like landscape I watered them myself. Today they are 30–40 feet high and include *dhok*, *chila*, banyan, and *sheesham*. As they grew the wildlife came—first the reptiles, especially snakes, then endless birds that could perch and make their homes in the trees; as the trees grew higher, peafowl, *langur* monkeys, *neelgai*, wild boar, and the occasional *chinkara* also started visiting. As the wildlife became richer on what was once totally barren landscape, the leopard and tiger were also occasionally seen. I remember in the early days watching with a torch a tiger and wild boar fight late in the evening, with the boar holding the tiger off. Even the occasional bear drifted in to eat fruit from the trees. Mongoose, jungle cat, and even caracal arrived. It just shows what is possible with a little protection and nurturing. That day I told Hamir this story of the regeneration of land while we watched a variety of birds coming to drink from a water pipe. I thought about the magical transformation that can take place on land just by planting trees.

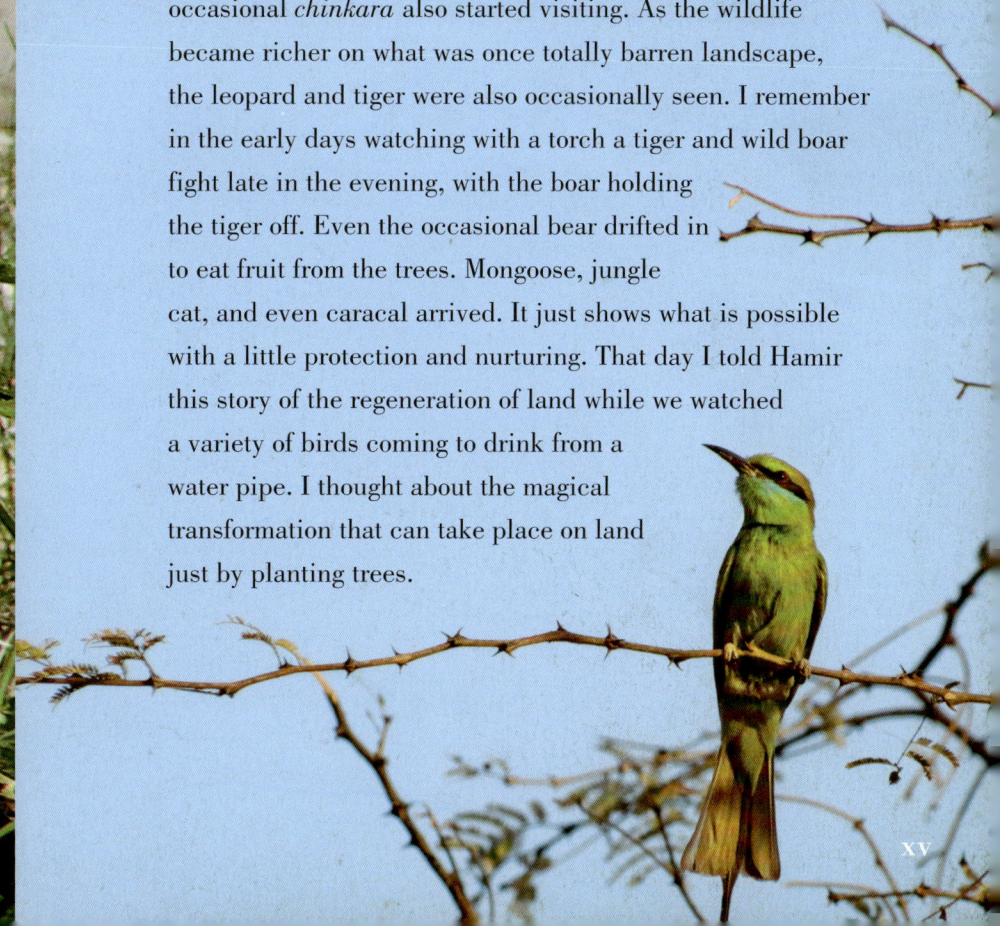

the
journey ...

Malik Talao

Padam Talao

Rajbagh
Palace

Jogi
Mahal

Ranthambhore Fort

day one

When we started at dawn on the first morning, we literally banged into four tigers the minute we entered the Park—a mother with her three 20-month-old cubs—at the edge of Rajbagh, one of the lakes hugging Ranthambhore fort. I did not realize then that tiger activity would be so intense that in all our nine days we would barely have a chance to leave the small 5 sq. km of area that encompasses the lakes and the fort.

That morning the family of tigers was feasting on a spotted deer in high grass. But the three cubs suddenly decided to roll out at 9:45am and they did what I call their *lake walk* across the edge of Rajbagh in typical Ranthambhore fashion. It took my breath away. One of the cubs started roaring and that awe-inspiring sound echoed across the valley. We listened in stunned silence. Our journey had started with a bang. Hamir clicked away with his camera as the tigers walked by. The scene was spectacular. Besides Hamir and myself, my nephew Jaisal and his wife Anjali had joined us; and soon Hamir's friends Ananya and Anoushka would arrive with their mother and grandmother, and so would our friends Raghu and Joanna. I mention all

this as my companions have given me some of their pictures for use; especially since I missed so much for the sheer pleasure of watching and I have no claims to being a photographer. When we left the Park at 10:25 that morning after watching the three tigers walk, I knew instinctively that something

special was brewing and Hamir and I would witness the old Ranthambhore magic. Jaisal had also been lucky with tigers in the past. He had managed to capture on film a tiger killing a *sambar* and a tiger fighting with a crocodile. The latter is the only record of such an encounter in the world. Maybe lady luck would ride with us.

For the last decade I have been saying to everybody that the 1980s in Ranthambhore had been another world which could never return. It's been difficult to explain to anyone who hadn't been there then. And I never believed that it could come back. I felt in my old bones that morning that the air had changed and crossed my fingers that the days that would unfold would be the experience of a lifetime.

When we returned that afternoon without Hamir who was sleeping, the three tigers were curled up in a corner 50 to 75 feet from each other. And what was remarkable was their alertness. It was as if something had triggered their activity. Suddenly one of them roared a few times. For me hearing tigers roar is soul stirring. I listened spellbound. It is always special to witness the final months that sub-adults spend with their mother. And this was a really special mother—Machli, a tigress I had known for fifteen years now—with her fifth litter. She had an upper and lower canine missing, but was still managing perfectly to cope with the enormous needs of her three cubs. Here in front of us was one dominant cub and two others that were more dependent on her. But it was clear that they roamed around freely, did what they wanted, and could also kill to eat. They were at an age when they endlessly marked their territory. I wondered whether we would witness something unforgettable.

Shyamji our driver told me that afternoon that the young Lakarda tigress was frequenting the lakes and sometimes there was much aggression between Machli and her. He said that the cubs also fought a lot and the large Lakarda male moved in and out of the lakes. This male had been collared so that his movements could be tracked. So from Shyamji it was clear that six tigers were roaming across the lakes. 'What bliss!' I thought. The finest landscape dotted with lakes and ruins of forts, palaces, chhatris, and masjids and tigers everywhere.

At 4:00 that afternoon, on the first day, the dominant cub rose and started walking. Amidst cars and jeeps, she

moved across the ruins between the lakes as the majestic fort loomed behind her. She went on marking whatever she could and the combination of tiger, lake, forts, and ruins gave me gooseflesh. Sheer beauty and the magnificence of nature mixing and merging with the memories of man— it is a unique combination that fills and overpowers the senses. Then suddenly a superb osprey flew off from above. I call this bird the tiger of the sky. It had just finished eating a fish. For most of that first afternoon we watched that dominant cub and one of her sisters as they moved back and forth between the fort and the two lakes. There was much roaring between them and the sound bounced off the ramparts of the fort. I sat listening in awed silence. When we came away just after 6:00 I knew that a special journey had begun. The evening reminded me of the nearly twelve family groups I had watched growing up on the lakes. And vivid scenes flashed across my mind—eight to nine tigers feeding on a blue bull at the edge of the second lake, all controlled by a female. What a day that had been! Then Noon and Nick Ear and the amazing exploits of Genghis who would chase and kill sambar deer in the waters of these lakes.

Back home I saw Hamir's pictures of the morning and he had got them right on! Steady and with good composition. They were the first pictures of his life, and those too of tigers. We had spent the day with Machli and her three cubs. And after eight hours of viewing I felt full of tigers deep within me. It triggered memories of when I had first

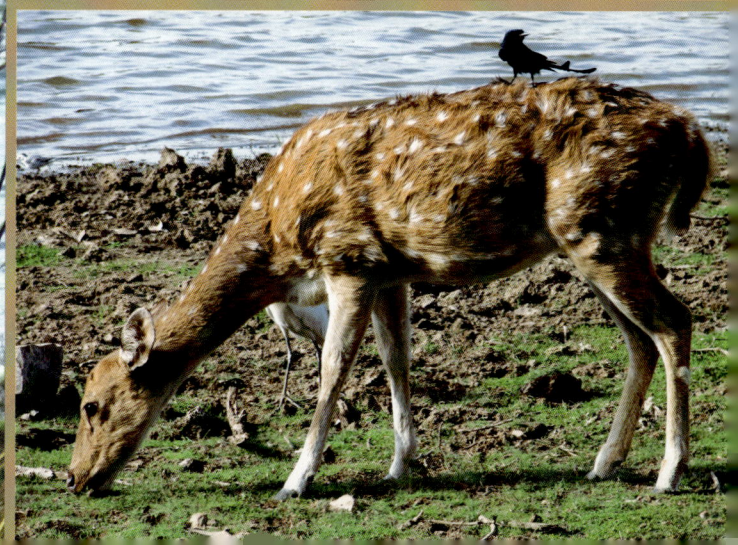

gone to Ranthambhore in 1975 at the age of 23. I still remember hiring a jeep and setting off with rations, including two crates of coke, to Jogi Mahal. Anyone could stay there, but few did. There were no visitors to the Park and tigers were near impossible to see. In the first years a sighting every few months was a reason for celebration and we moved around mostly at night. In summer we would sleep outside, close to the banyan tree. It was in these early years that Fateh Singh Rathore taught me about tigers. He was then completely engrossed in village relocation, the success of which led to Ranthambhore's tigers coming into their own many years later. But from 1975 to 1982, seeing a tiger was tough. Then excellent protection and a human-free environment created a tiger paradise. Today Ranthambhore has become a prime location on the planet for sighting wild tigers. It just shows what is possible with commitment and dedication. Simple strategies of protection and minimizing human disturbance can reap success. When I first visited Ranthambhore, there were twelve villages inside, all of which were relocated. Jogi Mahal was like home to me in those early years. Today the demands of tourism are so heavy that visitors are not even permitted to visit Jogi Mahal. Where there were no hotels or beds, today nearly 1000 beds exist for visitors outside the Park. The times have indeed changed. It is another world out there. Even our home, which was far from the madding crowd, is today surrounded by hotels.

Memories of time spent in the Park flashed by as I drifted off to sleep.

13

day two

By 6:30am the next morning, Hamir and I were being driven into the Park. The cool breeze of the March morning was delicious. And soon after, moving about the third lake—Mallik Talao—alarm calls from a bunch of spotted deer brought us to a grinding halt. We found, to our delight, a large male tiger known as the Lakarda male walking across the lake's edge. As he passed our jeep, Hamir and I both gasped in sheer wonder. That's the response of humans to the size of the male tiger. We stayed with him for fifteen minutes before he crossed into a ravine. He looked relaxed and well-fed. It all showed that the Park was being well looked after and patrolled. Nearly 200 ex-Army men had been hired to help with the protection. No grazing of livestock, no woodcutters, and no poachers meant the tigers were at ease. What a simple recipe to save wild tigers! The tragedy is that very few state governments have implemented even these basics. At least Rajasthan, over the last few years, had had a Chief Minister who cared and had created a special Core Group to supervise the protection of Ranthambhore and it was all working. I thank the Gods above!

As we came to Rajbagh we noticed a tiger in one of the windows of the palace—too far away for a

picture, but superbly positioned for binoculars. She looked like the dominant cub. Her two sisters were in the shade at the edge of the palace. I was sure the mother was sleeping in one of the rooms inside. What a scene! We watched for more than an hour before I decided to take Hamir to Jogi Mahal.

Hamir has childhood memories of Jogi Mahal, of the huge banyan tree and of climbing its endless branches with his mother. I have spent the best years of my life under this tree. It roots me to all that I love, and Jogi Mahal has been my home for so many years that to breathe the air around it is one of my greatest pleasures. We watched a troop of monkeys and endless birds on the lakeside while our forest tracker informed me that the tigress and her cubs frequented the tree. God, I thought, to see a tiger framed against its branches in daylight would be my thirty-three-year-old dream come true. Wishful thinking!

Hamir slept again that afternoon, 6am to 11am being a long stretch for him. I rushed back at 2:30 that afternoon to find all three cubs walking through the palace to a small islet before crossing over towards the fort. One went back and forth twice, and then the dominant one did her Ranthambhore walk, marking several trees and bushes on her way. She went across the masjid by the fort and lake and then to Chhoti Chhatri. Here she started roaring again, as if in constant communication with her siblings. Tiger chatter is a spellbinding sound that each day overwhelmed our senses more and more. The others cooled off in the shade and the afternoon was full of tigers in mind-blowing landscapes. Time rushed by as the tigers spent the afternoon between the lakes. My pictures tell the story.

day three

On the third morning Hamir and I gathered our equipment and were ready for the jeep even before it arrived. As I looked at the first light in the sky and sniffed the air, I knew it would be yet another remarkable day. I called it the morning of gate walks as we went through one gate following the dominant cub and landed up right near the entrance to the Park. What a shock to the men standing there! The tigress then walked along the fort's edge. Sitting down to groom herself, paws splayed, her tongue cleaned every regal inch of her body. I sat riveted. Hamir clicked away and got some superb portraits in one of the finest settings. In between, the same osprey was busy feasting on a fish. What a sight! It was one of my best encounters with this bird. Even the rare red spur fowl came out for a picture. My very first one in thirty-three years!

At one point the tigress paced towards our jeep and spotted a *cheetal*. In a second she had charged, whizzing past the jeep by 6 inches. Hamir and I were frozen by the speed and the sheer power of the moment. She then cooled off at the edge of the water amidst the backdrop of splendid monuments and fallen banyan trees. Ranthambhore has some of the most startlingly beautiful trees I have ever

seen. Suddenly she started roaring. Once more I was awestruck.

We returned from a morning full of wonder and beauty.

Hamir's friends had arrived, and early that afternoon we left him playing with them to go back to what we now called our three big cubs.

We found one sleeping in a bush near Chhoti Chhatri. She occasionally opened her eyes to watch the waterhole in front of her. It was 3:30 in the afternoon and very hot. The tigress watched as a group of cheetal started walking towards the water. A woolly-necked stork walked by. Four or five cheetal with a fawn approached the waterhole, some 40 feet from the tigress. Our jeep was on one side, another jeep on the other. The road separated the tigress from the deer. She was frozen, and completely alert. Then she stalked forward on her belly, moving about 4 feet from the bush. She watched intently, her tail flicking up and down. The cheetal quenched their thirst and started returning to the far side. But the fawn crossed towards us and 10 feet closer to the tigress. The tigress was now coiled to spring. As the fawn took two more steps forward, the tigress bunched up on her belly.

Suddenly she flew across the road in a blur, and four bounds later sat near the front tyre of our jeep choking the fawn to death. The spotted deer on the far side shrieked in alarm. An adult female paced down towards the tigress as if mourning the death of her fawn. She stamped her foot in alarm. The tigress rose carrying the cheetal in her canines and jumped into a pool of water. I could not believe it! She had killed a deer in front of our eyes and then taken it into a pool of water where she proceeded to toss it up and down like a football. She had glee written all over her face. It must have been one of her first few successful kills. What a sight! I rubbed my eyes in disbelief—in the middle of the afternoon I had seen a perfect natural kill and the tigress was in the water with her victim. …

It reminded me of the time when I had watched the tigress Noon attack a huge sambar stag in the middle of the day and try what she might, she just couldn't kill it. That event had provided me with one of the rarest sequences of pictures ever taken of a tiger's killing prowess. But this was nearly as good. She soon rose and carried the cheetal off to a bush where she ate for nearly one hour. Nothing was left when she emerged; even the bones had been

chewed up. She walked away, leaving us busily clicking our cameras.

On her way she roared loudly for more than a minute as if deeply satiated with her last couple of hours. Ranthambhore was giving of its best and rarest. I was moved to tears. What a scene to witness! And this carefree diurnal behaviour was made possible because of better protection and careful man management. In the last few years the Park Director, his assistant, and his rangers had been hand-picked, and the team in place had managed to work in concert. So had the local NGO that had, with this team, identified the poaching gangs. In this little microcosm of Ranthambhore something was working. What a scene, what an afternoon—a moment that would forever be etched in my memory. I came rushing home to tell Hamir of the afternoon's events. Jaisal and Anjali celebrated with some of their lovely pictures. I thought of Jaisal when he was 1–1½ years old, having a bath in a bucket in Jogi Mahal. He had also been linked to this amazing place for twenty-five years. Today he looked as excited as a child. That night sleep came slowly.

day four

Dawn on the fourth day was slightly overcast. Hamir was ready as usual at 6am with his two little girlfriends and a full jeep. In typical fashion, as we entered, we were greeted by two of the cubs. Hamir was now developing into an expert with his camera. I had to fight for space with him in the jeep! The tigers walked through the morning tourist traffic and ended up near the Chhoti Chhatri, whose architectural beauty in this dry deciduous landscape is a feast for the eyes. Carefully the tiger walked around the Chhatri towards Kawaldar at the corner of the fort. For me it was yet another first to get a backdrop of this Chhatri. We watched two of the cubs for nearly two hours, taking our eyes off them only to enjoy the beauty of a sparrow hawk and a pied kingfisher as it hovered over us looking for fish in a large pool of water. As it became warmer we went into Jogi Mahal and the shade of the banyan tree. The lake Padam Talao that faces Jogi Mahal is always full of birds and the odd deer in search of food. On the balcony of Jogi Mahal time stands still and peace, the like of which few know, envelopes the senses. Hamir and I paused for a cup of tea, and I recounted to him some of my old stories of Jogi Mahal.

That afternoon, leaving the children behind, we went back to our cubs. We found them at Kawaldar,

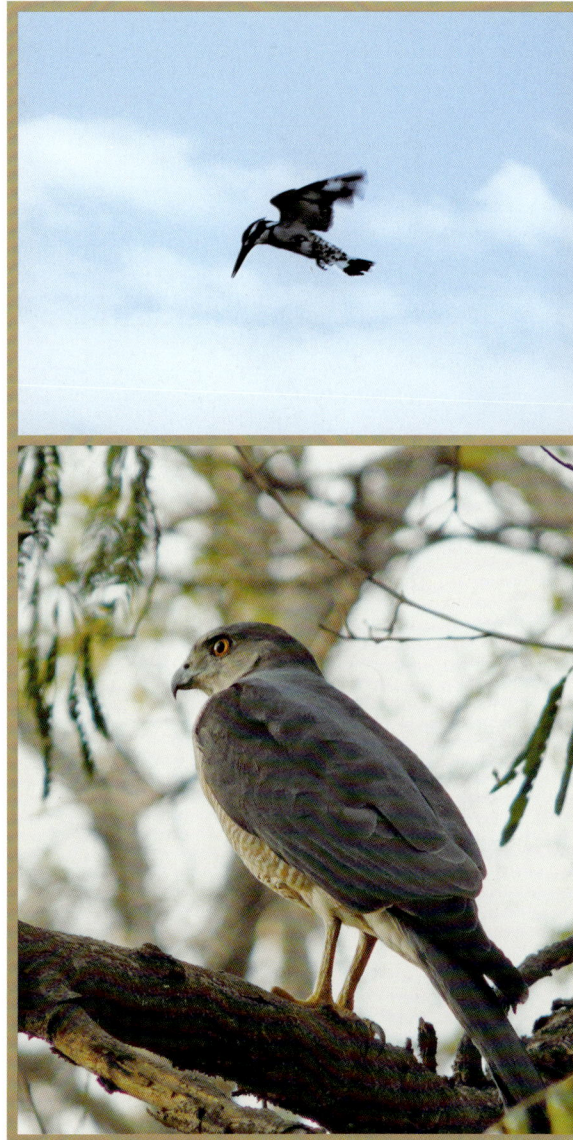

entering the Naal Ghati valley. The mother Machli and one of her cubs were resting in the *nalah* growling ferociously—I am sure that somewhere behind the bushes they must have gone for each other. Suddenly the forest trembled with the roars of the tigers. It was awesome, as their roars reverberated off the fort walls. Ten minutes later both tigers came out onto the road walking towards us, but an angry mother suddenly turned on her daughter and they rose on their hind legs to box each other with their paws. What a sight! I had only witnessed it once before between a male and female fighting over a kill. Obviously the mother had had enough of the cubs and was expressing her extreme irritation. I was certain she would leave them within weeks. They were a total nuisance since even though they could kill they demanded food from her. Today she had no patience whatsoever. They soon moved towards Naal Ghati, and twenty minutes down the road they passed a deserted labour camp full of pots and pans and human materials. Quaint sight! The cub sniffed its way past the camp and we then heard another jeep. Much to our surprise we found it following a third tiger. The occupant of this jeep was Ranger Daulat Singh whom I had known for

many years. We exchanged greetings as the tigers manoeuvred past us. But the cub Daulat had followed was now racing ahead to the first cub that was on the road one hundred yards ahead. Machli watched from the forest. This last cub drowned our conversation as the forest was split open by one of the most ferocious confrontations I have ever seen. The cubs were snarling, hissing, and growling, and with such intensity that it completely overwhelmed our senses. For five minutes the cubs boxed each other and we watched, awed by the power of the scene. When silence descended, I knew it was yet another first. By dusk all three tigers had retreated into the nearby nalah. I had witnessed an awe-inspiring encounter between two 20-month-old females. Our pictures show it all.

That evening Hamir and I pored over the pictures. Jaisal and Anjali joined us. Jaisal had got a superb picture of the fight. What a four days it had been! We sat satiated with tigers.

day five

On the fifth morning Hamir's friends had gone and the new passengers in our jeep were Raghu, one of India's leading wildlife scientists, with his wildlife photographer wife Joanna. That morning much spotting of birds of prey was done and by 9am we had once again found one of Machli's cubs near the lakeside. She walked across our jeep by the lake, veering towards Jogi Mahal. That is where we were going for a cup of coffee. We sat talking of old times as Hamir tried to click pictures of the banyan tree. Raghu had spent three months here in 1993 at a time when poachers had taken a toll and the tigers and Park were in total mess. The population of tigers at that time had crashed to 15.

Raghu's workplace was Panna National Park in central India where in the last few years the population of tigers had fallen from 34 to maybe 2 or 3, and all of them male. Though Raghu blew the whistle on the situation, the local government paid no heed. Sadly the simple interventions that had helped Ranthambhore in its recovery had been completely ignored in the neighbouring state. A tragic way to lose the tiger!

That afternoon we rediscovered our family between the lakes. But it was a quiet afternoon. We found a tigress asleep behind a bush at the edge of the lake.

A peacock alertly walked past her. The mix of colours and stripes was mesmerizing. Then, in typical Ranthambhore fashion, a cacophony of alarm calls caused us to race across to Rajbagh and, to our delight, we finally found the Lakarda tigress lying at the edge of the lake. We spent a great half an hour with her as she rolled on her

back and watched us. She was within hitting distance of the family and it was clear that two adult and three sub-adult females were circling each other amidst lots of aggression. The young and fit Lakarda tigress appeared to be getting ready to assert herself on the territory of the lakes. She soon rose and announced her presence by roaring for a minute, making sure that every tiger in the vicinity could hear her. Tiger sounds at dusk. This was tiger time and all our senses were overpowered by tigers—you could see tigers, hear tigers, and even smell them because they were so close. Maybe all this furious roaring was an effort to dislodge the 15–16-year-old Machli. But that would not happen without a fight. Even the cubs could join in since it would be in the interests of one of them to take over Machli's territory. The area of the lakes had become a site for competition, aggression, marking, and endless communication between tigers. Therefore all this non-stop tiger activity, with even the Lakarda male engaging with what might well be his future harem. I knew that Raghu had had a great first day. I saw his face beaming for the first time, after the serious depression caused by the fate of Panna's tigers in the last couple of years. And Joanna was thrilled with the sight of a bunch of parakeets attacking a

monitor lizard. One of the parakeets had even pecked the lizard's tail.

At dusk when we returned, we were all of us brimming with energy. Back home we told Hamir more stories of 'jungling'. By now every evening we would go out for a small night walk around my house. Hamir loves planets and stars, so each evening some night 'jungling' would take place. And as I gazed at the stars with Hamir I couldn't believe the time we were having. I thanked my lucky stars profusely. Looking at the stars was for me a moment of reflection. Thirty-three years after I first encountered its magic, Ranthambhore could still weave its spell around all those who were privileged to enter its portals. Hamir spotted Orion's belt and was in seventh heaven.

69

day six

A crystal clear morning but warmer than the others; I knew that summer was round the corner and as we moved past the entrance to the Park through the magnificent old gate I thought of how quickly temperatures would rise, within weeks climbing to 45°C. As we climbed the first hill Hamir shouted 'Ranthambhore'. It's that surreal moment when you first see the fort of Ranthambhore. It's a great sight, that first glimpse, for all Ranthambhorewallahs. That morning, yet again one of the cubs led us to the edge of Jogi Mahal and around the far side of the banyan tree. For a moment I sat with bated breath, hoping that my dream would come true and she would enter through the central branches. Alas, that did not happen. We found another of the cubs posing between the two lakes on a stony ledge as the early morning sun lit her and the walls of the fort. For me tiger and fort together are a sublime combination. She was soon joined by her sibling, and they both moved off into the area between the two lakes that faces the façade of the fort. They sat and calmly watched the hordes of visitors. Then suddenly, in a flash, they both raced towards Chhoti Chhatri as if called into the bush by their mother. What a morning it was going to be! Thirty minutes later one of the cubs came towards Badi Chhatri and

after walking around decided to move towards this massive structure. I couldn't believe it. I started to get ready to take yet another first-time picture of a tiger against this incredible backdrop. Lo and behold! She didn't just pass by but decided to leap into the Chhatri, and sat inside it watching us. What a splendid sight! I clicked away with Hamir till she felt the heat and moved into shadier forest. I was thrilled.

That afternoon when we returned we met the dominant cub near Chhoti Chhatri. At the edge of the forest the cubs had stalked a bunch of sambar deer. The sambar were frozen in fear and calling in alarm. But they were too big and clever for these two 20-month-old cubs to attack. Irritated, the tigress roared a few times before walking away. Her roars echoed off the fort walls. Time stood still. One of the cubs then moved past Chhoti Chhatri and once again, much to my delight, leapt into the Chhatri using it as a vantage point to gaze down at unsuspecting deer. It is amazing how the tigers of Ranthambhore have taken over the ruins. Instead of kings and queens, these 400-year-old ruins are now completely inhabited by tigers. It makes for incredible viewing. It had been another great day. I call it the day of the Chhatris! That evening my wife Sanjna arrived, and the tiger stories of the last six days reached a crescendo.

day seven

Every morning at dawn I would wonder if the incredible run with Ranthambhore's tigers would continue, but on the seventh day as well, soon after we entered, fresh pugmarks greeted us everywhere. And from the edge of Mori near the fort, we could see one of the cubs resting. Another cub lolled in the shade of a tree. We watched a couple of mongooses frolicking at the edge of the lake.

A couple of stone curlews with their huge yellow eyes shining brightly passed by the jeep. Looking up, I could see that the onslaught of the summer had forced the early flowering of the flame of the forest. A blossom-headed parakeet pecked at the flowers; a couple of rose-ringed parakeets joined in. We explored the far side of the lake, and watched the birds on its edges. Returning to Jogi Mahal we found a fresh set of pugmarks revealing the presence of what should be the third cub. As we climbed the steps to the balcony I found a small viper curled up on the table. In all these years I have never encountered snakes in Jogi Mahal. We made ourselves comfortable on the balcony and peace descended. We had taken clearance to spend the day there, and within an hour we spotted a tiger crossing the far embankment. It must have been 11:30 in the morning and in the hot sun it was walking across the lake. We watched spellbound.

The old times seemed to be back—tigers everywhere and coming out from all sides of the lake. I remembered how, nearly fifteen years ago, I had watched a tigress called Noon attacking four spotted deer on the edge of this lake. I was sitting on the balcony then and she killed one of the deer and the three others fled in total shock by swimming to the far side of the lake. That was their worst decision. None of them made it. The waiting crocodiles took them. The balcony of Jogi Mahal is silent witness to encounters that many of us have seen over the decades. Tigers have even walked in the lower balcony and kitchen. This is where we sat watching Genghis launch his attacks on both deer and crocodiles.

The tiger soon sat down to rest under the shade of a large tree on the edge of the lake. We quickly went across to within 10 feet of it, to watch it as it rested in that lovely landscape. Most of that splendid afternoon was spent between watching the tiger and looking at the antics of monkeys and at all the birds that visit the shallow waters of Padam Talao. The monkeys jump from the corner of Jogi Mahal straight into the tree, and there's always something that captures your attention.

At 3pm we moved towards the second lake and found both tigers resting. At the water's edge a painted stork tried desperately to catch fish. A long-legged stilt paced the edge of the lake, and in the distance an osprey made repeated efforts to catch a fish. By 4pm the first tiger started moving. It was a little cloudy and the sun was playing hide and seek with the clouds. The tiger crossed the masjid and the lake, much of its body silhouetted against an overcast sky. She then crossed towards us within a few feet of the jeep and continued down the road. The second tiger took exactly the same route, following its sibling. Utterly delicious light and landscape! What more could life offer, I wondered, as we spent the next hour with the tigers while they sat, groomed, and walked. Towards the end one of the tigers decided to call out. Another joined it. The roars of tigers filled the air. A superb end to the day. What a day it had been at Jogi Mahal and in the forest! The tiger across Jogi Mahal was still in the high grass, frightening both sambar and cheetal. Dusk fell and we left. We were thrilled that our tiger experience was continuing unabated. Even Sanjna, on her first day, had experienced the 'fullness' of tigers. As we went night 'jungling' with Hamir, I wondered what day eight held for us.

day eight

As we enter the Park we get news of tigers in Gular Kui, the lower reaches of the fort, and I spend some time debating whether to go there, knowing full well that much of the tourist traffic would be around the tiger. We do end up going only to find the mother Machli and one of her cubs above a waterhole. They look as if they have eaten and the remnants of their kill are in the grass. There are too many vehicles around so we move off, intending to return later. When we do, we find fewer vehicles and both tigers in the water.

The cub soon leaves the water and Machli continues to sit for twenty minutes, cooling off while her cub seems restless to move. Then both mother and cub start the long walk. They move nearly 4 km on the road to Jogi Mahal. For one hour we follow, luckily the first in a long convoy of vehicles. The cub claws trees, marks the ground, scampers, tries to play with her mother, and is in a much better mood than the other day when they boxed it out. It's all to do with having a full stomach. I see Daulat Singh, the Ranger, approaching from the far side. The tigers continue their walk, passing a gate with tigers painted on it in folk style. They manoeuvre past cars and traffic on the main tarmac road to the Park, passing by two stunned motorcycle riders. As they walk

upwards, the cub pauses to mark a milestone; it says Ranthambhore 1 km. Hamir watches bemused as mother and cub slide past our jeep. At the top of the hill, right below the ramparts of the fort, they pause and cross over the wall that leads to Jogi Mahal, which is also where we are supposed to have a cup of tea with the Park Director. It's been a superbly exciting morning and we get off at Jogi Mahal enveloped in contentment. We chat with the Director and Ranger on the balcony. Sanjna goes off to climb and sprawl on Hamir's favourite branch of the banyan tree. Peace descends, broken by occasional peacock alarm calls. It could be tigers a few hundred yards away. A few minutes later I hear the pattering of deer hooves, and as I turn towards the sound I see a herd of spotted deer scampering through the banyan tree; at the same time I see Sanjna racing to the steps of Jogi Mahal with Hamir in her arms. Time stops. I move a few steps forward, and then witness the most unbelievable sight as one of the tigers—the dominant cub—jumps into the tree just where Sanjna and Hamir had been seconds earlier. Sanjna used her instinct and fortunately moved away from danger. This created space for a tiger to jump in. I must say I was a bit nervous but there was no time to think. My dream had come true, and my camera clicked the first pictures ever in my life of a tiger under this ancient banyan tree; adrenalin flowed and the cub moved off. We were still excitedly discussing this amazing event when again, minutes later, a cheetal alarm call punctured our conversation and turned our attention back towards the tree where we saw the breathtaking sight of a tiger racing full speed under the tree after a cheetal fawn. I managed two shots before the tiger

vanished. Our excitement was at a peak. Hamir didn't know what had happened. I was again in seventh heaven, or is there an eighth one?

Only once in the last thirty-three years, sometime in the 1980s, on waking up in the middle of the night and looking towards the banyan tree from my window, I had caught a glimpse of a tiger slipping by. But I never really knew whether it had actually happened or if it was a dream. Right now I was near hysterical with excitement, telling the Park Director that it was a dream come true for me. He must have thought I was mad. Suddenly one of the jeep drivers said, 'Tiger coming back!' I rushed on foot to the corner of the tree even though the Director tried to stop me. I knew it was now or never. And I watched the tigress retrace her steps calmly and slowly, walking past the branches and roots of what has to be one of the finest trees of its kind in India. I carefully clicked, moving a few paces with her, a few steps at a time. My heart was pounding with

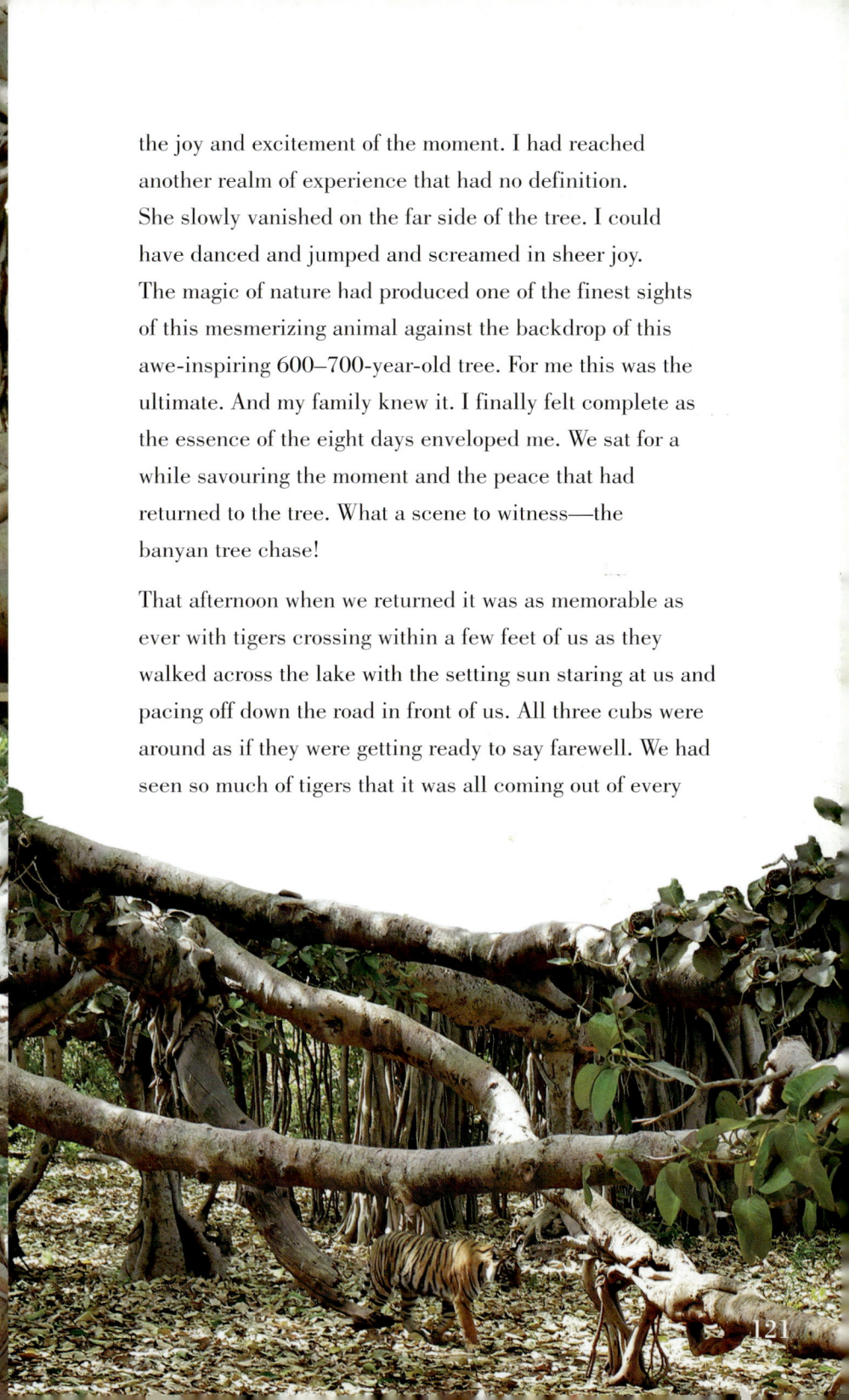

the joy and excitement of the moment. I had reached another realm of experience that had no definition. She slowly vanished on the far side of the tree. I could have danced and jumped and screamed in sheer joy. The magic of nature had produced one of the finest sights of this mesmerizing animal against the backdrop of this awe-inspiring 600–700-year-old tree. For me this was the ultimate. And my family knew it. I finally felt complete as the essence of the eight days enveloped me. We sat for a while savouring the moment and the peace that had returned to the tree. What a scene to witness—the banyan tree chase!

That afternoon when we returned it was as memorable as ever with tigers crossing within a few feet of us as they walked across the lake with the setting sun staring at us and pacing off down the road in front of us. All three cubs were around as if they were getting ready to say farewell. We had seen so much of tigers that it was all coming out of every

pore of our beings. And it was not just the sight but the sound as well with lots of roaring, growling, and snarling. The sight of tigers merged with the sounds and even the smell—that strong musky odour that comes when you get really close to tigers. We left two tigers on the edge of the road and returned home.

I don't think I have ever had such an amazing eight days, and today had been particularly incredible. Though Daulat Singh was still going to take us out on the ninth day, I knew that the trip was over and today had been its culmination. We did a little 'jungling' around that evening with Hamir and slept out in the open under the stars. Before I went to sleep that night my mind was racing with excitement. The unexpected had happened. I had now to put picture to paper. I had to bring out the essence of this very special window of time that Ranthambhore and its tigers had allowed me, and that too at a juncture when the tiger was facing its worst crisis in India.

125

day nine

Three of us went with Daulat Singh for what was a very quiet ninth day. It had drizzled at night after a huge storm, and the forest fragrance of wet earth was delicious. Most junglewallahs delight in it. Strangely, once again my jungle instincts seemed to be working. I asked the driver to turn the jeep upwards beyond Rajbagh, and Shyamji suddenly said, 'Tiger!' There were no alarm calls, and in total silence the Lakarda tigress walked out of the forest and passed within hair's breadth of our jeep. As silently as she came she went back. There was not a sound of alarm from anywhere. I had to rub my eyes to believe that what I had seen was really a tiger. It was a special farewell to us. And the very best kind you can get on this planet.

I met the Director and Ranger before catching the afternoon train. Deep down, I was moved to tears by the pleasure this sojourn had given me. I told the Director that Ranthambhore was living a moment like when at the peak of its success in the late 1980s. For myself I had found a little bit of hope and some optimism that maybe some tiger homes would survive irrespective of the lack of political will. I told the forest staff who must have seen my exuberance as a change from my gloomy depression of the last few years that this was one of the greatest moments for Ranthambhore in the last twenty-five years. And it was a moment that I had thought would never come. In a way Hamir and I, and even Sanjna over the last few days, had seen in reality the stuff that dreams are made of. May the tigers of Ranthambhore keep creating dreams for the world to see!

postscript

Sanjna wanted to return to Ranthambhore and we did so on 12 and 13 April over a long weekend. Obviously her two and a half days had not been enough and I didn't mind. It would give us an idea of whether the tigers continued on their magical course in Ranthambhore. It was much hotter but again, as soon as we entered the Park on the first morning, we found one of the cubs cooling off in a back pool of water in Rajbagh. Another one seemed to be nibbling something in the grass. They then went towards the palace looking satiated and lay in the shade of a tree at the edge of the palace entrance. A third one also lay nearby. These Palace tigers are amazing, and that afternoon I saw one of the cubs entering the Palace and later on two of the

tigers were immersed in the lake waters at its edge. These tigers haunted the Palace just like ghosts from the past. I was reminded of the story of a sound recordist late one evening hearing music and laughter from the Palace but being unable to record the sounds as his machine could not register them!

Lots of tigers but a little lethargic because of the heat, and therefore more in the water. The other cub was around the masjid and Chhoti Chhatri, and even the next morning we watched one of the cubs leap into the Chhoti Chhatri from where it looked out like the monarch of all it surveyed. I could feel the 150 visitors in their different vehicles ready to applaud. What a stage and

what a performance! She later went into the forest and surfaced near Jogi Mahal on the far side of the banyan tree. In this short visit we also had news of another tigress with her two 6–7-month-old cubs, who were on a sambar kill some 10 km from the lakes. So we took time off to go and catch a glimpse of this new family which was much more evasive. The cubs were delightful to watch, well concealed and far from the noisy and intrusive jeeps.

Amazingly, while the tigress ate on one sambar kill with her cubs, another sambar came to the waterhole and the tigress ended up making a second kill. Mother and cubs spent over a week eating the sambars in the most gorgeous of settings.

That final afternoon on the tarmac road to Jogi Mahal, Sanjna said, 'Tiger!' It was her moment. I least expected it. But there was one of the cubs coming straight towards us. After about 100 m, she suddenly changed her mind and moved towards Jogi Mahal again. As she was marking the 1 km Ranthambhore milestone yet again, a motorcycle carrying two adults and four children came from the other side—six people on one motorbike. For a moment tiger, the motorbike passengers, and we were immobilized. Then in a flash the man leapt off his bike and started running backwards with his family. He was shouting, 'Bring my keys. …' The tiger watched; I was amazed and stunned as little children scampering on the road 50 feet away didn't affect the tigress at all. She continued on her way, but our worry about the children forced us to push her a little to the side. The tigress could easily have pounced on the small children. I remember once on the same road a child had gone towards a bush to relieve himself at dusk, and had been attacked and killed by a tiger, the only incident of its kind in Ranthambhore's history. But it was amazing how every nuance of human movement has been assessed by Ranthambhore tigers in the friendliest of ways. She walked

on, pushing the traffic upwards. Cars reversed, traffic was in a mess, and the odd person ran back including the motorbike rider who was coming back to try and retrieve his motorbike. Suddenly she branched off and moved towards the fort. In the 10 minutes that she went up, the frantically worried man had managed to hitch a lift to his motorcycle and ride it back to where his fearful family was waiting. The tigress came back down the road and paced onwards. Lo and behold! The motorcycle rider and his family were coming back around the corner and confronted the tigress yet again. But this time he had room to swerve and he carried them back from where he had come. I noticed that his seat cover had a design of fake tiger stripes! We were in splits of laughter in the jeep. Where else on earth could you witness a scene like this? More soberly, where else would you find a wild tiger watching a

family of six from a distance of only a few feet away! And it went to show the remarkable gentleness of the tiger. A gentleness that humans lack. I knew that for tigers to live freely we must give them their space and not allow human beings to overrun their territory. Tigers and human beings cannot coexist and all our policies to save wild tigers must be based on this fact. Ranthambhore's wonderful tiger activity was a result of the massive village relocation programme in the 1970s. I remember one night in 1977 when village Lahpur was about to move. I wasn't sure then that it would have an impact. But thirty-one years later I know that this was the key to the success of Ranthambhore and all because of one man, Fateh Singh Rathore, who put his heart and soul into the relocation effort and then managed to put Ranthambhore onto the world map. In the middle of the traffic jam, the tigress jumped over the wall and went across to Jogi Mahal. This is when a fast asleep Hamir woke up and said, 'Tiger'. We had carried him asleep into the jeep when we left.

The cub had proceeded to walk around the far side of Padam Talao, and we watched the tiger yet again from Jogi Mahal's balcony.

Sanjna and Hamir soon left to look at the new family 10 km away. I stayed alone on this last evening clicking some beautiful backlit portraits of the two cubs, the first having been joined by a second at Mori. They were again heading towards the Palace for the evening. The sky had turned deep grey and a storm was brewing. The breeze started whistling. I was certain the cubs would be watching this storm from the Palace balcony. The sun was dipping and this was my final farewell. But I would be back …

Hamir's scrapbook

photo credits

Valmik Thapar: Cover, ii, vi, xii, xiii, xiv, xv, 2, 4, 5, 6 (top right, bottom right), 7, 8, 9, 10 , 11, 14, 16, 18, 19, 20, 21, 22, 23, 24 (bottom right), 25, 26, 27, 28, 30, 31, 32, 35, 36, 38, 40, 41, 42, 43, 44, 45, 46, 48, 49, 50 (top), 53, 54 (left), 55 (right), 62, 64, 65, 68, 70, 71, 72, 73 (top left), 76, 77, 78, 79, 80, 81, 82, 84 (right), 85, 88, 89, 91 (top right, bottom right), 92, 93 (right), 95 (top left), 96, 98, 100, 102, 105, 106, 108 (right), 112, 113, 115 (right), 116, 117, 118, 119, 120, 121 (top left, bottom left), 122, 123, 124, 125, 126, 127, 129, 130, 131, 132, 133, 134 (left), 135 (right), 136, 137 (top left), 139 (left), 143, 148, 152, Back Cover

Sanjna Kapoor: vii, viii, 24 (left), 67, 73 (bottom left, top right, bottom right), 93 (left), 95 (bottom left, right), 99, 108 (bottom left), 111, 121 (right), 137 (right), 138, 139 (right), 140, 141, 142 (left), 144, 145 (right)

Hamir Thapar: 150, 151, Jacket Back Flap

Raghu Chundawat and Joanna Van Gruisen: v, 6 (left), 12, 17, 24 (top right), 60, 61, 66, 69, 74, 83 (right), 84 (left), 86, 90, 94, 101, 107, 108 (top left), 109, 134 (right)

Jaisal and Anjali Singh: 33, 34, 37, 39, 103, 110, 114, 115 (left), 142 (right), 145 (left), 146, 147

Amita Chabbra: 50 (bottom), 51, 52, 54 (right), 56, 57, 58, 59

Daulat Singh: iii, 104

Debbie Banks: x

Sanctuary Photo Library - Jagdeep Rajput: End Paper